CLIMATE OF ECOPOLITICS

CLIMATE OF ECOPOLITICS

A Citizen's Guide

First Edition

**Written
by
Paul Taylor
Environmental Scientist**

iUniverse, Inc.
New York Bloomington Shanghai

Climate of Ecopolitics
A Citizen's Guide

iUniverse books may be ordered through booksellers or by contacting:

iUniverse
1663 Liberty Drive
Bloomington, IN 47403
www.iuniverse.com
1-800-Authors (1-800-288-4677)

Because of the dynamic nature of the Internet, any Web addresses or links contained in this book may have changed since publication and may no longer be valid.

ISBN: 978-0-595-50152-6 (pbk)
ISBN: 978-0-595-61393-9 (ebk)

Printed in the United States of America

To my darling wife Daryn Plancher—
Whose patience and loving attention sustain me daily.

CONTENTS

Preface

This work of non-fiction is an overview assessment of 21st Century environmentalism where *ecopolitics* have replaced rational scientific discovery and discourse in the hysterics surrounding the theories of global climate change. Global warming has been identified as both the world's greatest crisis, and the world's greatest hoax—neither is true.

The term *environment* essentially means *surroundings* (36). Unfortunately, its meaning has been stretched, conflated, contorted, misappropriated, abused, bastardized and politicized to the point of trivia by activists and media to manipulate public policy in a range of issues that are as endless in their scope as often misguided in their ends. Environmentalists have become skilled at gaming the government regulatory systems for political advantage in the guise of progressive public service as tax-exempt organizations. This is today's ecopolitics.

Today, computer models can generate compelling scenarios for any political argument on an untestable proposition about a future hypothetical environmental threat. Cloaked in a veneer of pseudo-science, a hypothesis often sustains positions of environmental activists where a tenuous scheme of worst-case assumptions requires the rational observer to prove the irrational negative proposition. Here, the obscure, often immeasurable environmental impact can be promoted as an imminent threat. This *fear mongering* has become the routine and systematic, yet disingenuous tactic to erect counterfeit public issues for global political exploitation (9).

The reader is invited to consume this work with a mind open to the new idea that global environmentalism has devolved into a partisan political special interest in the 21st Century. Sadly, consensus science that solves real environmental problems has been replaced with pernicious political propaganda and demagoguery. And, the cost-benefit analyses necessary for prioritizing and solving environmental problems goes unmentioned in the issues of global climate change—unmentioned because such analyses are incalculable in today's knowledge of climate science.

This book attempts to condense the massive, and often contradictory, volumes of information on global warming and climate change for citizen consumers. This book cites several of the contemporary and credible experts on the issues of climate science and climate policy for a balanced assessment.

Acknowledgement

The publication of this book has reasonably applied *best efforts* in citing and referencing sources consulted in its preparation pursuant to the *fair use doctrine*. The author wishes to express his sincere gratitude to the consulted sources in adding to the body of public knowledge on subject matter as important, dynamic and often intractable as *the environment*.

Introduction

Few, if any, public issues have the personal resonance of environmental issues. Each of us can feel both victim and perpetrator in environmental issues. Environmental issues have become an integral part of world culture, and for some, a personal moral cause. The environmental movement has grown worldwide to become the largest, most densely organized political cause in human history (9). And, the global environmental regulatory systems, begun in earnest in the 20th Century, have often erred on the side of caution under the paternalistic maxim of "prudent avoidance" without complete science for the actual cause-and-effect of environmental impacts. Every personal and professional activity in your daily life involves some environmental regulation. Now, global warming theorists claim its cause and effect in every other known environmental issue—global warming has become environmentalism's universal scapegoat and obsession. Over 600 global maladies have been attributed to global warming including such preposterous things as a Minneapolis bridge collapse and the Katrina disaster, and the decline of circumcision, brain size, cremations and Earth's rotation (2). A complete list of these wildly speculative and often silly items is contained in Chapter 5 herein.

Predictably, environmentalists identify the sinister enemies of the environment as greedy corporations, property owners and the prosperous lifestyles of developed democracies. It starts with media access that promotes civic fear and misapprehension about environmental issues, and points to capitalist democracies as conspirators against you and our environment. Here, environmental activists become vital media assets, perceived as heroic *saviors* of the planet. This is the *Alice-in-Wonderland* area where the possible crisis becomes the probable, the suggestive becomes the conclusive, the accusation becomes evidence. The long-term cost-effective solutions to environmental problems derive from moderate centrist policies that do not attract radical activists, crisis-hungry media or political extremes.

From a broader perspective, and without a clear scientific consensus on the validity of such things as global warming, activists have hijacked political agendas in climatic doomsday scenarios that, if true, are the most apocalyptic so far. However, if false, the trendy global warming hysteria may be the final blow in the precipitous slide in legitimacy of environmental activism. Ironically, in the 1960s and 70s, environmentalists successfully defeated the expansion of nuclear power plants in the US that could have substantially reduced greenhouse gas-producing fossil fuel use impacts which those environmentalists now blame for global warming. Nuclear power plants emit no greenhouse gases—zero *carbon footprint*. Nuclear power plants also reduce US dependence upon foreign oil (9).

There is a profound arrogance in the belief, without firm scientific proof, that human activity can significantly impact global climatic conditions. 95% of the sources of the predominant global warming gas carbon dioxide, which comprises only 0.03% of our atmosphere, are naturally occurring sources (3).

Neither politics nor legal systems have been accountable consistently to the truths of environmental science. Environmental laws and regulations of any kind have the effect of incrementally limiting personal liberty and freedom, and of imposing severe, and often unnecessary, financial burdens on those who can least afford them.

When dealing with attitudes toward the Earth's ecosystem, we must recognize the absence of a commonly accepted philosophical base for the development of an environmental law system. As a nation, we have no controlling ethical relationship with our natural environment; some persons wish to exploit, others to preserve, and most are in between. As a result, the basic framework of an environmental law system develops slowly, often with inconsistent provisions on the same subject. And this development of law occurs in a political setting that is often acrimonious, for the basic values of some members of our society are challenged by new laws dealing with the environment.

But as national attitudes change, so do the laws. For example, wilderness in biblical references was the antithesis of a garden, and gardens were created by man conquering nature. To the Puritans, wilderness was a hostile environment, a last refuge for sinners. Wilderness in 18th and 19th Century western literature had an evil connotation. Often the phrase "howling wilderness" was used. Today, we have a Wilderness Act, Wilderness Areas, a Wilderness Society, and wilderness conferences—all more or less in praise of wilderness. Other concepts relating to the environment have been subject to similar changes of attitude.

An example of these attitudes can be seen in what we today generally regard as a national treasure—the Grand Canyon. Yet the first western man to see the canyon, Don Lopez Cardenas, did not even write of his journey. The first report, by Pedro de Castaneda, gives a vivid account of the difficulties of the descent into the canyon, but not a word of its beauty or grandure. Three hundred years later, US Army Lt. Joseph Ives explored the canyon. He concluded that the Colorado River at the canyon's bottom would be an economical avenue for the transport of supplies, but that the area was a profitless locality which would forever be unvisited and undisturbed. Ten years later, John Wesley Powell, the first man to follow the river through the entire canyon, described it as the "Grand Canyon." In 1903, Theodore Roosevelt called the canyon a natural wonder unparalleled throughout the world, and sought to protect it. Thus, the history of the Grand Canyon is an example of the changing attitudes of man toward nature.

Nature was once thought to be the enemy against whom man struggled to survive. Later, nature became the source of exploitable natural resources and wealth. In recent time we have begun to appreciate and analyze the relationship of man with nature, and man's interdependence with nature. This "newer" theory calls for man to utilize nature's bounty, but to exist so that the dynamic natural system survives, hence *sustainability*.

The historical development of environmental laws and regulations parallels the evolving philosophical views of nature. Prior to the beginning of the 20th Century, the legal system was used to encourage the development and exploitation of our natural environment through land use rights. The government encouraged highways, railroads and canal building, and gave away millions of acres of public lands to those who would exploit them. Beginning in the late 19th Century, the concept of conservation became popular, and this resulted in legislation in the early part of the 20th Century to conserve and manage our natural resources.

Conservation means many things to many people, and shortly after the conservation movement began, it split into two camps or schools of thought—a dichotomy that exists today. John Muir (American naturalist, 1838–1914) led the protectionist or nonexploitation conservationists, while Gifford Pinchott (American forester and politician, 1865–1946) was the leader of the "careful extraction" or "wise-use" school of conservation.

The basic principles of conservation have long been expressed in many hundred of ways—well before conservation became popular. Some aspects of the ages-old nature religions were conservationist at heart; others caused vast ecological destruction in the worship of nature. For example, the Teotihuacan people of Mexico's Classic period destroyed their forests to make lime ash for stuccoing their huge "City of the Gods." This contributed to the drop in the water table and the drying of their corn fields—all while worshipping various nature gods and goddesses, including the goddess of water (9).

Conservation is a way of life. And, it should be made clear that living in harmony with nature need not require the giving up of many of the advances that a thousand years of applied science and technology provide human civilization. Science has to do with proving physical facts and relationships using the *scientific method*. Science requires the exclusion of human emotions and political preferences to discover truth. Science is the antithesis of humanism.

CHAPTER 1

▼

GREEN HISTORY

The environmental movement emerged in the early 20th Century as a collection of conservationists, naturalists and bird watchers. The movement grew slowly until government began recording natural resource impacts of human activity, such as over hunting, timber clear cuts and strip mining. In the last half of the 20th Century the movement exploded worldwide on government recognition of the potential health effects of environmental transgressions. Today, one is considered to be uncivilized if unconcerned about the environment. Today, environmental matters are a free-for-all of global political pomposity and propaganda.

It has been said that the original environmentalist was the US's 26th President, Teddy Roosevelt, who established the National Park Service in the early 1900s to set aside wilderness lands and conservation areas. During his 1901 to 1909 Republican presidency, Roosevelt designated 150 national forests, the first 51 federal bird sanctuaries, 5 national parks, the first 18 national monuments, the first 4 national game preserves and the first 21 land reclamation projects. He placed 230 million acres of land under federal protection. Teddy Roosevelt was the first to use the word "conservation" in describing US government policy. Roosevelt had come to the view that "the irresponsible use of natural resources is a fundamental problem which underlies almost every other problem of our national life."

With two World Wars and The Depression intervening, it would take nearly sixty years to complete a federal government framework for conservation. Initially, the concern was smoke, sewage, and such. The clean air, clean water, and solid waste acts of the 1960s were still animated mainly by aesthetic insults or the appearance of being dirty, offensive or untidy. As with conservation, the principal objectives were still visible, tangible and use-oriented. Clean air, swimmable water, and the proper containment of solid waste dumps were emphasized. For example, New York City isn't the Grand Canyon, but the clarity of its air is still worth conserving, if only because a city can be a scenic asset. The Endangered Species Act, passed unanimously by the Senate in 1973, seemed to come from the same old conservationist perspective. Cougars, bears, buffalo, and wolves were to be protected. Teddy Roosevelt would surely have voted to protect these animals from extinction as eagerly as he would have hunted them when they were abundant. These laws extended the simple conservationist philosophy to its logical, practical and measurable limits.

US Government is formally committed to encouraging "productive and enjoyable harmony between man and his environment." This promise was made in the National Environmental Policy Act of 1969 (NEPA). Passage of the NEPA in 1969 was an attempt to create a new frame of reference for the consideration of all major activities by the Federal Government: a frame of reference that would include consideration of impacts to the environment. NEPA was the culmination of a decade of previously unsuccessful Congressional attempts to define and put into practice a national environmental policy. The Resources and Conservation Act, proposed in 1959, called for the Executive branch to coordinate its scattered conservation efforts. The Ecological Resources and Surveys Bill of 1966 contained provisions designed to remedy the inadequate use of environmental data by federal agencies.

NEPA had a three-fold purpose: to establish national environmental policy, to authorize research concerning natural resources, and to establish a council of environmental advisors. Later Federal action required agencies to prepare an "environmental impact statement" if any of their proposed actions might significantly impact the environment. Another purpose for enacting NEPA was to fulfill the need for an interdisciplinary approach to environmental management and decision-making in all branches and levels of the Federal Government. NEPA provided new approaches for dealing with environmental problems on a preventative and anticipatory basis, and represents a break from the previous practices of dealing only with environmental crises and attempting to reclaim resources from past abuse.

Even as the Federal Government completed the regulatory framework for traditional conservation, these laws also quietly launched the new era of environmentalism to control pollutants. To begin with, regulating pollutants of any kind requires a more elaborate and intrusive regulatory force than regulating the more human-scale parks, wilderness lands and wildlife. In 1970, President Nixon established a new cabinet-level agency under the US Department of the Interior, that would be the US Environmental Protection Agency (EPA). The EPA was established to take charge of all US pollution control and associated regulatory programs. More significantly thereafter, each of the new laws also included something quite new—an open-ended "toxics" provision, a general invitation for the EPA to monitor remote and invisible environments for hazardous pollutants and regulate them as needed for public health and habitat protection. In addition, though written with cougars and wilderness mainly in mind, the Endangered Species Act of 1973 (ESA) had been similarly expanded broadly enough to protect such exotics as the flower-loving fly, the fairy shrimp and other critters whose actual numbers in their natural habitats can never be counted. The ESA would soon be expanded to cover "habitat modifications" as well.

A mere statutory afterthought in the 1960s, remote and invisible environmental pollutants get entire Federal Laws of their own a decade later. The

Toxic Substances Control Act is implemented in 1976. Then the Superfund hazardous site cleanup program in 1980, followed by RCRA, CERCLA and SARA and other ubiquitous acronyms for government regulatory expansion to monitor and control every detectable impurity, which technology now allows us to identify at concentrations of less than one part per billion. For example, think of a football stadium filled to the very top with white marbles, with the ability to detect a single red marble among them, and then speculate with computer model projections that the single red marble at some concentration, at some distant, future time, is a hazardous pollutant that should be controlled by the Federal Government. You are reminded that the EPA began with a budget of $1 billion and 4,000 employees in 1970. Today, the EPA's budget is about $8 billion with 18,000 employees (4). This does not include the thousands of contract regulatory researchers operating under the EPA-administered grants whose millions of dollars will never be accounted for. This US growth and enforcement in environmental regulations has largely been replicated in other western democracies until today where the United Nations is leading the global concern over climate change and global warming.

Somewhere between the Vietnam War and the Love Canal property contamination episode near Niagara Falls, the legal infrastructure of the "new environmentalism" slips into place. Conservation isn't abandoned—it is just overtaken politically, subsumed into something bigger. Bigger, paradoxically, because it concerns the very small, the remote, the invisible yet detectable chemical agent whose environmental impact, quantifiable or not, can become the fear of the moment for environmentalists and media exploitation.

Environmentalism, and its environmentalist believers, didn't become a potent public political movement until the 1960s and 70s in the US when college campuses, brimming with idealistic baby boomers, were determined to make every new emotional twitch a political movement—a cause for revolution. This is when political movements, valid or not, became television news programming assets, and when anti-establishment and counter cultural influences became media partners in a way that is largely taken for granted today. Today, news and entertainment mass media have become indistinguishable and readily exploitable as the propaganda machine in the game of ecopolitics (24). In today's ecopolitics, environmentalists convene as media events. Politicians arrive to say "they care" to placate the environmentalists for short-term media and electoral rewards, enabling the irrational regulatory infrastructure to expand without end or accountability to measurable scientific cause and effect. Environmental policymakers have little patience for conclusive scientific evidence. Environmental policy is now achieved through regulatory fiat to sate political activists who themselves can no longer be bothered with letting scientific rigor get in the way of what they want. Here is a tragic legacy of the 1960s and 70s cultural revolution; where the "rhetoric of virtue" and well-intentioned compassion-baiting dialogue make truth a negotiable commodity to advance one's public political ideology (9).

Environmentalists and their groups do not make their reputations or raise funds by making public pronouncements about the enormous world progress in solving environmental problems. What environmentalists don't want you to know is that all human activities come under environmental regulation, and that most pollution problems are solved, or are under active management.

CHAPTER 2

▼

ENVIRONMENTALISM

Environmentalism has member believers whose daily lives involve a conscious self-reflection about human uses, conditions, consumptions, products, ambitions, ideology and survival as negative impacts upon natural systems. This self-reflective being, with rational to wildly irrational motivations, is an environmentalist.

Environmentalists

Who are environmentalists? What do they believe? What do they want? Well, demographically, they're mostly middle to upper-middle class white, college-educated, agnostic or un-religious, and seemingly well-intentioned people with lots of time on their hands. Most have never taken on the responsibilities of parenthood or running a business. They generally rely on academic and government institutions for their wisdom and values, and prefer group identity over self-reliance. Environmentalists have a romantic fixation on simplicity and even passivity as rules for living in a complex and energetic world. Ironically, simplicity and passivity make life poorer, not greener.

Radical environmentalists are intolerant of growth, prosperity and free enterprise. These eco-freaks rabidly resist measuring their goals against other critical concerns, like economics, national security and positive results. In their extremes, they constitute an "axis of antagonism" that provides no product or service for future global prosperity. They reflexively detest competition, capitalism, political diversity and corporate globalism.

There are three basic and sustaining, yet fallacious, public perceptions that today's movement environmentalists (the *ecologista*) nurture and exploit. These public perceptions—indeed, misperceptions—persist because valid environmental analysis ultimately reduces to science, not feelings of compassion, guilt, fear, narcissism or political opportunism. The public and its media have little interest in, or understanding of, science, and are easily manipulated by anyone with the facility for the hysterics that promote controversy and insecurity among the citizenry.

First, there is the flawed public perception that all local or regional environmental impacts, irrespective of their relative scale or ecology, are somehow cumulative to, and compounded with, all other known environmental impacts to result in a catastrophic, irreversible global environmental collapse. There simply is no empirical example of any such terminal collapse occurring or likely to occur as a result of human activities. The expanding ripples on 19th Century writer/educator/preservationist Henry David Thoreau's *Walden Pond* were contained succinctly within the pond water, without impact to air or land. Even if one of the three natural resources of land, air or water is impacted, its condition does not necessarily add to, or compound with, other impacted resources in any grand global cataclysm. Provocative speculation that all individual impacts to the natural resources of land, air and water result in significant, irreparable ecological impairments is simplistically and functionally flawed.

Second, there is the public misperception that all living ecological systems are inherently fragile and mysteriously complex. Living ecological systems are powerful and wondrous forces whose resilience and adaptability far exceed man's relatively brief tenure and influence on Earth. Man has a tendency to conceitedly view all living systems in human consciousness—to anthropomorphize nature. Unlike man, nature acts only in the elegant efficiency of survival without ideology, morality, economics, politics, psychology, compassion or historical reflection. Nature is both intrinsically dominant and infallible—man is neither. Ecological systems have evolved exquisite assimilative capacities, mutations, bioremediations, dispersions, redundancies, regeneration, recovery and energy management capabilities. Ecological systems are not inherently fragile. The contrary is true. The scientific record of life on Earth demonstrates comprehensively and conclusively that ecological systems can and will sustain themselves infinitely in time, form and among all natural resources. This is true for atmospheric constituents associated with climate change.

Third, there is the public perception (again, misperception) that business and industry enterprises have a vested sinister interest and covert mission to destroy and pollute natural resources without regard for ecology or human health. Business and industry are motivated by profit, not pollution. Profit and pollution are ultimately incompatible goals for long term business and industry survival in 21st Century regulatory systems. Legitimate business and industry have adopted environmental compliance as part of their routine production and public relations objectives. Business and industry have demonstrated over the last 30 years that they will address environmental issues as long as regulatory controls are applied consistently among their competition, and as long as regulations are based in rational science with measurable positive results. Long established and enforced regulations that limit industrial air pollutants coincidentally reduce greenhouse gases associated with global warming (9).

Prime Movers

An original scholar in the theories of man's conflict with the sustainability of natural resources (the environment) was Thomas R. Malthus (1766–1834). Malthus was an English clergyman and economist. Malthus published an anonymous pamphlet in 1798 promoting the theory that human population increases geometrically, while food supplies can only increase arithmetically. Therefore, sooner or later, the growing gap between food supply and food demand must end in war, famine and general human misery. Malthus simply argued that when mankind reaches the limits of nature—when it had farmed all the farmable land—mankind would starve. With the technology of 1798, Malthusian theory was both obvious and true in postulating that the ascent of man causes the collapse of everything else and that, in turn, destroys man too. Now, over 200 years later, clearly Malthus was in error because he grossly underestimated man's ever-evolving ingenuity. Human resourcefulness was left out of the equation.

Born and raised on a farm in Springdale, Pennsylvania, Rachel Carson wrote a seminal book on the overuse of US pesticides in agriculture titled *Silent Spring* in 1962. DDT and its derivative chlorinated hydrocarbons were agricultural pesticides developed in the late 1930s. DDT was widely used in farming to improve crop yields, and was found to control almost any type of insect. These chlorinated hydrocarbons are quite resilient in the environment, having decomposition half-lifes of 10 to 15 years. Though not an ecologist, Carson's book implicates reckless worldwide pesticide use in the unraveling of ecosystem food chains in very personal, almost romantic language. Carson predicted that pesticide use would eventually contaminate world drinking water supplies. DDT use in the US was banned by Federal Law in 1972. Carson's prediction, fortunately, was wrong (9).

A more consequential environmental scholar, probably because he wrote during the 1960s and 70s era when college campus political movements *flowered*, is Stanford University biologist Paul Ehrlich. In 1968, Ehrlich wrote a bestseller titled *The Population Bomb*. Ehrlich's book embraces and extends the Malthusian theory toward the irrational. Ehrlich's 1968 doomsday vision stated, "The battle to feed all of humanity is over. In the 1970s and 1980s, hundreds of millions of people will starve to death in spite of any crash programs embarked upon now." Time has proven Ehrlich to be flamboyantly and profoundly wrong. He admonished that nature will take its revenge against mankind's abuses. Further, Ehrlich attributes the AIDS epidemic to the "deterioration of the epidemiological environment which is quite directly related to population size as well as to poverty and environmental deterioration." Ehrlich appears to have stumbled from his basic life science expertise into elitist psycho-socio-babble, to his professional discredit (9).

US Senator, Vice President and presidential aspirant Al Gore authored a 400-page tome on everything environmental titled *Earth in the Balance: Ecology and the Human Spirit*, which became a "national bestseller" (42). Gore, in writing the book, said he was willing to risk his entire political career on the issue of *the environment*. His original 1990 title for the book was "World War III"—ironic for a 2007 recipient of the *Nobel Peace Prize*. Gore's emphasis was that to attain his version of global environmental rectitude would require the commitment and sacrifice of world war. *Earth in the Balance* rambles maniacally among premises of over population, nature's spirit, technophobia, consumptionism, species extinction, and introduced us to the term *global warming* in an attempt to circumscribe an "environmental holocaust without precedent," and to position Gore as the political leader whose insight will save the planet.

Gore exposes himself as either profoundly confused, or cynically manipulative, about the meaning of "technology." He writes "[G]overnment, as a tool used to achieve social and political organization, may be considered a technology, and in that sense self-government is one of the most sophisticated technologies ever created." His further abstractions equate technology with "spoken language," and even "the human body." Gore also elaborates on how technology is not necessarily science. One should always be suspicious when politicians begin to bend the meaning of words. For clarity, please observe that Webster's Dictionary defines technology as "applied science" (37). Gore also calls for science and religion to be "reunited in the service of the environment."

Gore, in keeping with the world war analogy of his environmental crusade, promotes vast government programs such as a "Strategic Environmental Initiative" and "Global Marshall Plan." His Global Marshall Plan would act to stabilize world population, develop environmentally appropriate technology, measure environmental impacts in economic terms, develop international environmental regulatory programs and develop a global environmental education program. His Strategic Environmental Initiative was named to imply an environmental equivalent of the "Strategic Defense Initiative" (SDI), the crash program to develop a series of technological breakthroughs focusing on a common military objective, which Gore opposed as senator. Gore's Strategic Environmental Initiative would be a global "program that would discourage and phase out older, inappropriate technologies and at the same time develop and disseminate a new genera-tion of sophisticated and environmentally benign substitutes." This, from a politician who clearly does not understand the definition of the word "technology."

Gore further writes to introduce the divisive environmental concepts of "environmental justice" and "sustainable development" in pandering to the *ecologista*. These latent environmentalist concepts are attempts to lever-age social issues of immigration, class warfare, racism and big business bashing for pure political patronage. Gore climaxes his pretentious and partisan policy masturbations with a renewed self-assurance in big govern-ment dominance over personal ambition, liberty and free enterprise.

In 2007, Al Gore received an *Oscar* for the movie documentary *An Incon-venient Truth* that was a cinematic version of his slideshow tour, wherein he conceitedly claims that "… the debate about global warming is over." Gore's "truths" have been widely discredited by climate scientists from whom Gore has never accepted numerous challenges for public climate debates. Gore, far from ending the debate, actually started the debate about global warming. *Vanity Fair,* magazine of social record, validation and wisdom for our plutocrats, had a "Green Issue" that attempted to rally eco-consciousness with the fanciful cover headline—"A Threat Greater Than Terrorism: Global Warming." Ironically, this headline is both a denial of the horrors of *9/11*, and an admission that the US is winning the war on terror. The magazine presents Al Gore, among other *green elites*, as the moral arbiter and oracle on all things environmental—including global warming.

A more recent example of the endless desire to identify with *things environmental* is the pseudo-scientific, psycho-social, co-optation called *ecopsychology*. Every political movement has its psychological dimension. This is no less true for the environmental movement. Persuading people to alter their behavior always involves probing personal motivations and values. Political activism begins with asking what makes people tick. What does the public want, fear and care about? How do we get and hold the public interest? How much can people take; what are their priorities? Have activists overloaded the populace with anxiety and guilt? Ecopsychology endeavors to address the problem of effective communication with the general public in order to meet the demands of the "environmental revolution." Ecopsychology claims to "redefine sanity within an environmental context," to re-examine the human psyche as an integral part of "the web of nature" or ecosystem. Ecopsychology presumes to "bring together the sensitivity of psycho-therapists, the expertise of ecologists and the ethical energy of environmental activists" (38). Here is a perfect example of the abstraction of a physical science discipline (i.e., environmental science) with a soft pseudo—science discipline (i.e., psychology, sociology, etc.). This is sophistry, not science, and certainly no basis for making prudent management decisions about our natural resources or climate change.

The fuzzy regulatory area of non-, or not-for-, profit tax-exempt organizations deserves scrutiny as regards the global environmentalism and its groups. Most of the well-known environmental groups operate as tax-exempt organizations. This tax-exempt status means that nonprofits are effectively subsidized by all other taxpayers. Austensively, tax-exempt status infers that these groups are taxpayer subsidized because they act in the *public interest* to enhance and protect *public welfare*, and they therefore deserve the *public trust*.

According to the US Internal Revenue Service (IRS), general categories of tax-exempt organizations include those organized and operated exclusively for one or more of the following purposes: religious, charitable, scientific, testing for public safety, literary, educational, protection of children or animals, or amateur sports promotion. Few of the dominant tax-exempt environmental organizations can be seen to strictly qualify under one or more of the foregoing IRS categories. Many of the prominent eco-non-profit organizations frequently jeopardize their tax-exempt status under the IRS "propaganda" prohibitions. The IRS Code states, "… (organization is exempted provided) no substantial part of the (organization's) activities … is carrying on propaganda or otherwise attempting to influence legislation or intervene in any political campaign on behalf of any candidate for public office." Propaganda is defined as ideas, facts or allegations spread deliberately to further a cause or to damage an opposing cause. Eco-nonprofits have been cautioned by the IRS for lobbying and propaganda activities that exceed the "no substantial part of activities" tax exempt status standard under the IRS Code (9).

In 2004, the IRS began increased oversight of the approximately 1.3 million nonprofit organizations operating in the US under IRS Code Section 501(c)(3). Thousands of charitable organizations routinely endorse political candidates in local, state and federal elections. These endorsements also come from environmental, civil rights, labor and other "research and education" nonprofits. It is therefore not surprising that one-third of US adults think that the nonprofit sector is headed in the wrong direction, according to a Spring 2005 Harris Poll. Only 10% of donors believe that charities are honest and ethical in the use of donations. Nonprofits have grown out of control and unaudited over the last thirty years, coincident with political polarity in America—environmental nonprofit groups grew from 2000 to 4000 during the 1990s. One survey found that only 144 of the 200,000 nonprofits founded in America since 1970 has reached more than $50 million in annual revenue (7).

Over the last 30 years, ecopropagandists have done a good job of frightening and shaking the general populace into first the hysteria of "the sky is falling," then more recently into resentment over the exaggerated daily claims of environmental apocalypse, until today where warning-battered rational people are barely listening to environmentalists. After years of being bombarded by ever more dire ecological prophecies, of which none has materialized, citizens have grown more and more skeptical of environmentalist predictions and protests. The alarmist theories that environmentalist made fashionable during debates over acid rain, toxic groundwater, nuclear winter, over-population and species extinction, have backfired. Citizens have now adopted a "selective deafness" as a first line of defense against wild claims of environmental disaster (9).

Ecopsychologist Theodore Roszak noted in his 1993 book, *The Voice of the Earth*, that the environmental movement might have overutilized shame-and-blame tactics in its approach to the public. And, that the public may become particularly vulnerable to right-wing conservative attempts to instigate a "green backlash." Roszak theorizes that the green backlash may provide people an opportunity to avoid feelings of guilt and helplessness, and to attack environmentalists who make them feel that way. Roszak's bizarre musings reveal environmentalisms perverse and cynical manipulative devices of promoting public guilt and helplessness (i.e., *victimization*).

As the focus of environmentalists moves from once-immediate dangers now under control, to more abstract matters of aesthetics or sustainability or global warming, latent economic class conflicts are beginning to erupt. Pollution controls often impose highly regressive costs according to socio-economic class. For example, in the early 1990s, the cost burdens of Southern California's aggressive air quality management plans were estimated to have a three times greater impact upon the region's poorest households than on the wealthiest. Environmentalists dismiss such economic irritants by arguing that a better environment helps everyone. The proverbial ecological crisis notwithstanding, the adverse health consequences of reduced economic opportunities for the poor vastly overwhelm any environmental benefits they may enjoy from, say, marginally cleaner air quality. US air pollution reduces average life expectancy by approximately 30 days. Poverty strips away 10 years in life expectancy.

The experience of 30 years of environmental controls in the US is testament to conserved natural resources and solved pollution problems, and conclusively demonstrates that growth-oriented economies (i.e., free-market democracies) actually do a better job of managing natural resources than a society run on the myopic principles and utopian-directed theories of environmentalist dogma (9).

Green Is the New God

Environmentalists are blindly resentful of the broad growth in economic opportunities that comes from free enterprise. Environmentalists see themselves as heroic (if not messianic) figures in a movement that is more socialistic than problem solving. This is where the environmental ideology can take on the trappings of a religious crusade. The movement transcends the need for truth to aid proselytization.

Global warming has spawned fanatic believers in abstract theories, mythologies and mysticism in today's environmental movement that reach a near fetishistic fervor. Sadly, the productive scientific motives and technologies that solve environmental problems have been bypassed and replaced by perverse political dogma, demagoguery and fearmongering propaganda. The *green* movement began in prospering 20th-Century democracies such as the US and Western Europe; because their affluence absorbed the costs of environmental controls—today about 5% of US GDP. The environmental movement grew to become the largest, most densely organized public cause in human history. Today the movement increasingly mimics religion in its identification of humans as *sinners* against nature, its calls for personal redemption via lifestyle changes, its claims that environmental issues are now moral issues, and, with the predicted global warming apocalypse, it has its biblically-proportioned *Armageddon*. As with other religions, environmentalism also has its *false prophets*, and its extremist followers in the form of animal rights cults and *eco-terrorist* activists.

The religion of environmentalism has a simple, godless good-versus-evil orthodoxy: nature does good, man does bad. This basic belief enables anti-capitalist, *internationalist*, and global-socialist political operatives. The faithful congregations of environmentalism are found under the roofs of the United Nations, the European Union, the global Green Party, the Democrat Party in America, and the over four thousand nonprofit eco-groups. These organizations are the dogmatists for the eco-religious— environmentalism is their *de facto established religion*. Further, these are the institutions from which environmentalism's dogma and false prophets exploit political opportunity. Eco-activist's claim moral authority in requiring your adherence to the cause of environmentalism. Your personal commitment and sacrifice for their fanciful moral imperatives such as *global sustainability, smart* _____ (you fill in the blank), *carbon footprinting* and *environmental justice* provide your path to environmental enlightenment, purity and ultimate redemption. Many of these feel-good goals have become common policy language in local, state, national and international environmental regulations. These policy goals are fed wholesale to our children in *politically-correct* public school curricula where the mention of traditional religions has been banned for decades—kids are programmed to *green worship*. Global news media have been a willing and gullible accomplice in the hysterics of environmentalism, and in particular, regarding global warming.

As with all religions, environmentalism has the harmful impacts of false prophets, cult extremists, and conflicts with proven science. No where in the history of the environmental movement have the false prophets and extremists been more duplicitous than in the eco-propaganda and political exploitation of the theory of global warming. Wild doomsday speculation from green group fund raising propaganda have distorted the issues of climate change far beyond rational scientific discovery or discourse. Recent published science concludes that the Earth is warming, but that the crushing global economic impact of fully-implemented Kyoto Protocols would only reduce global warming gases by less than one percent during the 21st Century. Contrary to religion, science is not about belief, hyped hypotheses, corporate conspiracies or political opportunism. The applied sciences that enable and protect your everyday activities are about repeatable, measurable proof of a theory for cause and effect concerning physical phenomenon using the *scientific method*. Policymakers must resist environmentalism's seduction of short-term political gains, and await the conclusive science that properly analyses global climate change. Short-term leverage of environmentalism for political expediency will only hasten and lengthen a global economic collapse, with far greater human suffering, conflict and pollution than any of the worst-case scenarios being proliferated by environmentalism's 21st-Century climate crusade.

CHAPTER 3

▼

ECOPROPAGANDA

Politics and truth seldom occupy the same space anymore. When the environmental movement found a political base, it began to leave the truths of scientific rigor in natural resource management policy behind as too slow, too cumbersome. In the climate of ecopolitics, the environmental movement has lost its way and legitimacy in the 21st Century. Environmentalism has become just another partisan means to a political end.

Trans-Science

In order to glean some truth from any activist representations, one must start from the cautious position that the first casualty of activism is the truth—truth is secondary. Activist, including environmentalists, rhetoric is tactically reliant upon exaggerated dangers and inflammatory word use to erect counterfeit arguments and promote bumper sticker platitudes as moral authority for social engineering. Typically, the *zero sum* false dilemma proposed is that of "it's only man or nature" that will survive the current environmental disaster (23). Anyone with only a rudimentary knowledge of the living systems of Planet Earth, that include both man and nature, understands that nature, with or without man, shall ultimately prevail just as it has for over 3 billion years prior to man's unceremonious and primitive appearance on Earth about 3 million years ago.

The founding beliefs of environmentalists are that of scarcity, of limits to growth and therefore, "the sky is falling," "the end is near," "catastrophe is just around the corner" fear-mongering proclamations. Their basic and transparently political theory for the collapse of the Earth's ecosystem is blamed mainly on the democratic-capitalist way of life. Their *guilt game* carries the further shame that, for example, Americans consume 28% of the worlds natural gas, 23% of the solid fuels, 20% of the coal, 23% of the crude oil, 42% of the gasoline, 26% of electricity, and 10 to 30% of its copper, aluminum and zinc, and drive more and larger vehicles more miles than any other nation (31). One can only surmise that the operative environmentalist theory is that the high quality of democratic-capitalistic life must somehow be a threat to the Planet Earth.

Where is the truth? Recent scholars in the true environmental sciences have coined the term *trans-science* to describe the study of phenomena that are too large, too diffuse, too rare, too distant or too long term to be resolved by reliable scientific methods. This trans-science defines the current boundary of understanding climate science and global warming. Exploiting trans-science, environmentalists release so-called "studies" that bypass the scientific journals and peer review, and go straight to sympathetic, issue-hungry and largely gullible journalists (33). The due diligence of true scientific cause-and-effect findings is neglected.

Of course, without each of us applying some critical examination and context to these waves of scary news, the complex, remote and invisible threat will make you uneasy as it is designed to do. It is also designed to create civic anxiety, demonize business and industry, and thus promote a dependency on government to solve a problem that may not exist in the priorities of our lives or global sustainability. Consensus in legitimate science has always led to rational truth about physical phenomena such as climate change. Today, the radical environmental movement is more alchemy than altruism. Today, the media is more concerned with the controversy that attaches to environmentalism, rather than truth telling about the state of the environment.

American courtrooms and government regulatory agencies have been overwhelmed by health scares linked to environmental issues—pesticides, ozone depletion, electromagnetic waves from cell phones, etc. Too often, government environmental researchers operate as a self-selected, insular, academic group with an inherent bias toward identifying environmental risks. Often, this leads them to become personally, emotionally and financially wedded to their own theories, and scientific objectivity is lost along with the truth (34).

Today, a successful government regulatory scientist may find possible environmental problems, and then publicize them as probable environmental crises to get the attention of legislators for follow-up work and renewed funding (19). Often the incentives in government are to save, rather than solve the problem, and thereby save bureaucratic power and it's vast taxpayer-supported *civil service* employment opportunities and its massive partisan voting blocks (17). Today, there is an unholy alliance of environmentalists, media and regulatory bureaucrats worldwide that conjures up environmental evils, and dresses them up as science in a system that is perfectly evolved to fund and grow the global government establishment. There are legions of academics and regulatory scientists whose occupation it is to invade, critique, punish, and ultimately dictate your lifestyle—ecopolitics reigns.

Green Marketers and Psychologies

Green marketers have developed sophisticated schemes to sell an avalanche of eco-friendly and sustainable products. They believe that consumers will support green products, when given the right information. The marketing themes, as with for other products, must emphasize an immediate, practical and emotionally-compelling benefit in a way that relates to the daily lives of the target buyer. Green marketers use detailed research survey data in an attempt to develop hard-hitting messages that make environmental protection tangible and relevant. They also deploy these messages through highly-leveraged partnerships with other products, institutions and media that are already a part of the consumer's lives. But, while 86% of Americans express concern about *the environment*, only 1% see it as the most important problem, only 44% call themselves "environmentalists," only 12% have voted against a candidate for environmental positions, and only 26% have bought a product for environmental reasons (8).

One of the more speculative *green products* is associated with global warming insurance marketing schemes. Global insurance companies are finding new marketing opportunities in the assault on global warming and climate change. The following initiatives are under consideration:

- Travelers, the giant insurance firm, will offer owners of hybrid cars in California a 10 percent discount. It already offers the discount in 41 other states and has cornered a large share of the market;

- Fireman's Fund will cut premiums for "green" buildings that save energy and emit fewer greenhouse gases. When it pays off claims, it will require environmentally-friendly products to replace roofs, windows, and water heaters, etc.;

- Marsh, the largest insurance broker in the US, will offer a program with Yale University to teach corporate board members about their fiduciary responsibility to manage exposure to climate change.

The insurance industry's clout is sizable. It's the second-largest industry in the world in terms of assets, and has a direct link to most homeowners and businesses. It insures coal-fired power plants as well as wind farms, so it can influence the power industry's cost structure. With its financial muscle, the industry can require the use of new financial instruments designed for companies to trade greenhouse-gas emissions in the same way that commodities are bought and sold.

The insurance industry has the ability to change behavior, costs and policies, and to squeeze millions of clients. Some consumers are already noticing a negative effect in their insurance. In the past year, some 600,000 homeowners living in a zone that an insurer considers a high storm risk in an era of climate change have seen their policies cancelled or not renewed (44). While the insurance giants are echoing *green alarmists*, they are also attempting to cut coverages and raise premiums on individuals and business by advanced speculation on the uncertain and incomplete scientific cause and effect of global warming.

There has been a profusion in the marketing of books on *the environment* over the last 30 years. Unlike this book, they predominately cover "the sky is falling" mantra, and a call to guilt-driven personal sacrifice and recruitment to organize in the revolutionary cause of environmentalism. An interesting benchmark of environmental activism is the histrionic doomsday propaganda that filled books released to coincide with the twentieth anniversary of Earth Day on April 22, 1990. This date is a generation removed from the original Earth Day established by Republican President Nixon in 1970. The following referenced excerpts are examples of the hyperbolic claims made by authors of the environmental books that sought to capitalize on the book sales genre of Earth Day 1990. The subsequent Al Gore book *Earth in the Balance* is analyzed in preceding chapters, and should be considered in a class with the following books.

- "The Exxon Valdez. Alar-treated apples. Love Canal. Bhopal. Three Mile Island. Chernobyl. The list of individual examples of environmental degradation could fill this book. But none of these calamities should be thought of as isolated incidents. Cumulatively, they add up to the worst state of environmental affairs since the dawn of civilization. And they've contributed to a fraternity of staggering problems whose scientific jargon has become the common language for consumers as well as scientists: global warming and the deterioration of the ozone layer, acid rain and urban smog, deforestation, toxic waste, garbage overload, water pollution." (*Save Our Planet*, Dell Trade Paperback, 1990.)

- "Clean water is our most precious resource. But as much as a fourth of the world's reliable water supply could be rendered unsafe for use by the year 2000." (*Save Our Planet*, Dell Trade Paperback, 1990.)

- "The Amazon basin, 3.1 million square miles of it, has the world's largest and most biologically diverse tropical forest. It contains anywhere from one-tenth to one-half of the planet's plant, insect, and animal species, depending on who is making the estimate. Edward Wilson of Harvard University calculates that forest destruction worldwide causes the extinction of about 10,000 species every year. In addition, one-quarter of all our medicinal drugs are derived from tropical forest plants. These include medicines that treat heart disease and childhood leukemia. We will never know how many other valuable drugs have been lost by forest destruction." (*Our Earth, Ourselves*, Bantam Books, 1990.)

- "Fifty acres of rainforest are destroyed each minute. That's almost 27 million acres a year, an area equal in size to the state of Pennsylvania. At no time in history has the rate of deforestation approached what we are seeing as we enter the 1990s. Two-fifths of the world's original rainforest cover has been decimated, mostly in the last fifty years." (*Save Our Plant*, Dell Trade Paperback, 1990.)

- "The Nature Conservancy says extinctions are accelerating worldwide. Our planet is now losing up to three species per day. That figure is predicted to be three species per hour in scarcely a decade. By the year 2000, 20% of all Earth's species could be lost forever." (*50 Simple Things You Can Do to Save the Earth*, Earthworks Press, 1989.)

- "Eighty percent of America's solid waste is being dumped into 6,000 landfills, spread across every state in the country. But that option is shrinking fast. In the past five years, 3,000 dumps have been closed, and by 1993, some 2,000 more will be jammed to capacity and closed. In just four years, Chicago's landfills will be full; dumps in Los Angeles should reach capacity by 1995." (*Save Our Planet*, Dell Trade Paperback, 1990.)

- "The Environmental Protection Agency estimates that in the next five to ten years more than twenty-seven states and half of the country's cities will run out of landfill space. Major cities including New York and Los Angeles will exhaust their landfill space in just a few years." (*Earth Right*, Prima Publishing, 1990.)

- "When the federal Superfund Law was enacted in 1980, many people hoped that the federal cleanup program could be a short-term, one-time effort. It now appears that the task of cleaning up hazardous waste sites will haunt us well into the twenty-first century. Not only is it taking longer to clean up sites, but new sites continue to be discovered. Over 300,000 locations now contain hazardous substances, and the number is growing at a pace that far outstrips the rate of cleanup." (*Save Our Planet*, Dell Trade Paperback, 1990.)

- "Some sixty US counties, including much of the urban Midwest and East, violate minimal air quality standards, spewing more pollution into the air than is legally permitted under the federal Clean Air Act. The American Lung Association believes that about 115 million Americans are being exposed to treacherous air pollution levels. The American Academy of Pediatrics believes that as many as 28 million children have been put at risk because the air is too dirty to breathe safely." (*Save Our Planet*, Dell Trade Paperback, 1990.)

- "Even if the entire world were to stop using CFCs and halons immediately, destruction of the ozone layer would go on for decades.... The destruction of the ozone layer over Antarctica from CFCs and halons used today will continue into the twenty-first century. The Natural Resources Defense Council estimates that even with an immediate total ban on ozone-depleting chemicals, recovery of the ozone layer will take more than a century." (*Earth Right*, Prima Publishing, 1990.)

- "In 1985, a 'hole' was found eating its way across the sky above Antarctica. It is now believed that this hole is as deep as Mount Everest is tall and as wide as the United States. Most scientific researchers are convinced that global CFC emissions must be reduced substantially, if not completely, to avoid a catastrophic depletion of the stratospheric ozone layer." (*Save Our Planet*, Dell Trade Paperback, 1990.)

Note the over-hyped language use such as: "calamities," "worst state of environmental affairs since the dawn of civilization," "staggering problems," "unsafe," "forest destruction," "destroyed each minute," "decimated," "species could be lost forever," "option is shrinking fast," "jammed to capacity," "will run out of landfill space," "waste sites will haunt us," "spewing more pollution," "treacherous air pollution," "28 million children put at risk," "ozone hole found eating its way across the sky," "catastrophic depletion." This dramatic language sounds more like Hollywood horror movie advertising tag lines than enlightenment about managing our natural resources. Needless to say, none of the preceding eco-horrors ever occurred or is likely to occur (9).

Psychiatrists, psychologists and sociologists have long postulated that symptoms of irrational fear and anxiety increase when political and economic systems are most stable and prosperous. This is why institutionalized enviro-mythmaking has proved to be quite powerful, persuasive and pernicious. The involved institutions are mass media news sources such as TV, radio, the internet, magazines and newspapers. The basic, raw material for mass media is controversy, to feed the activity of environmental fear mongering. University of Southern California Professor of Sociology, Barry Glassner, wrote a seminal and best-selling book on fear mongering in 1999 (39). Glassner debunks (among others) the 1990s enviro-myth of the dangers of US schools containing asbestos. Government regulators estimated that one-third of the nation's schools contained asbestos insulation that when inhaled over long periods of time can cause lung cancer. That school kids would be exposed to the asbestos health risk, became a public outrage. US schools spent over $10 billion to remove school asbestos even though its removal posed a greater health hazard risk than allowing the asbestos to remain installed and immobile. In this case, media engaged relentlessly in the school asbestos health scare as fear mongering; as they have with other health scares such as AIDS, Dow Corning silicon breast implants, Gulf War Syndrome, road rage and now global warming. Why do fear mongering media campaigns take hold? Why do media and their audiences get drawn to one hazard over another?

Fear mongering motivates people to 1) correct a moral offense or 2) to criticize a disliked group or institution (40). Health hazards, at any degree of injury or prevalence in the population, are deemed to be morally unacceptable, whether merely perceived or real in scientific medical terms. Witness the school asbestos example above as a motivating environmental health hazard. Some of this motivation arises out of a collective cultural narcissism or sense of entitlement; because real or not, health risks are innately personal and have resonance with our basic survival instincts.

The second fear mongering motivation to criticize or discredit a disliked group or institution came into full influence in western culture during the 1960s and 70s. Then until now, demonizing groups or institutions has become the activists' sport and even occupation to leverage environmentalism's distrust of corporate free enterprise. Given the high threshold for motivating moral outrage and the seeking of personal redemption via public political protest that have characterized environmentalism, they are not likely to be motivated by an aforementioned moral offense in response to environmental issues. Rather, the large, faceless target is big business. Some of this antipathy has its roots in the historic labor union conflicts with big business and pure Marxist socialism.

The isolated, dramatic, personal anecdote of some environmental issue is the *smoke* that is fanned into the *flames* of public outrage by media for government to regulate big business, be it industrial manufacturing, mining, oil, home builders or their financiers. With the flames lit, the more public talk there is about the reported environmental issue, the more likely are other accident-monitoring agencies such as police and insurance to collect similar examples of the reported issues that they would have ignored altogether or classified differently prior to the reports. Psychologists call this the "Pygmalion effect." Further, fear mongering relies upon what psychol-

ogists refer to as the *availability heuristic*, where people judge the significance of an isolated issue by how readily it comes to mind. When we are
presented with a survey that polls the relative significance of an issue, we
are likely to give greatest significance to whatever the media emphasizes at
the moment because that issue tends to come to mind (39).

Fear mongers make their scares all the more credible by having professional spokespersons or *victim-cum-experts* to spread pseudo-scientific or
dramatic testimonial information about an environmental issue. Professional narrators also play an important role in transforming implausible
environmental threats into the "disaster de jour." The rantings of alarmist
newscasters and the glorification of wannabe experts are two tricks that
expose the fear mongers manipulative ploy. In addition, the use of tragic
anecdotes in place of statistically-tested scientific evidence, and the confla-
tion of isolated incidents into trends or conspiracies are an attempt to
exploit deeper cultural anxieties and hatreds, and even personal paranoia.
The simple answer as to why we are exposed to so many persistent and
irrational fears is that immense political power and financial rewards await
those who tap into our moral insecurities and supply us with symbolic
substitutes (39).

The US wastes tens of billions of dollars and human resources every year
on fear monger-promoted enviro-mythological issues, including research
and technology, and on victim compensations for *metaphorical illnesses*.
Metaphorical illnesses include Gulf War Syndrome, multiple chemical
sensitivity and silicon breast implant disorders where people justify their
personal fears, prejudices, hardships and political ideologies in the absence
of determinative scientific explanations by projecting themselves into a
public class of victims and litigants (40).

Metaphorical illnesses have become popular in recent personal and class
action injury lawsuits. Here, gullible, sympathetic and big business-loath-
ing juries and jurists have evolved a system that perversely exhibits the
wealth redistributive characteristics of Marxist socialism. Take for example

the hundreds of billions of dollars involved in state health agency *litigation jackpots* won by damage suits against the big tobacco companies.

Cause and effect findings of environmental impacts can be reliably established when the significance of valid scientific data are subjected to statistical analysis—it tells you whether sufficient quality and quantity of data are available to determine the cause and effect theory tested. Scientific cause and effect cannot be determined by coincidence, personal anecdotes, circumstantial evidence or preponderance of court evidence which are prevalent legal standards, not scientific conclusions. Scientific cause and effect has not yet been developed for accurate projection or remediation of the possible impacts of global warming.

CHAPTER 4

▼

CLIMATE SCIENCE

Neither politics nor legal systems have been accountable consistently to the truths of environmental science. Government regulations of any kind have the effect of incrementally limiting personal liberty and freedom, and of imposing severe additional costs on those who can least afford it. Therefore, the establishment of scientific cause and effect is a critical prerequisite to prudent environmental regulations.

Scientific Method

Science has to do with proving physical facts and relationships using the *scientific method*. Science requires the exclusion of human emotions and political preferences to discover truth, and science is the antithesis of humanism. The scientific method involves a number of basic principles that one should engage consistently and systematically. These principles should be applied in an orderly manner with the appropriate technique, helping to ascertain an answer to a question. Scientific method is a process that allows new knowledge of natural or physical phenomena to be acquired, gaining explanations by sifting the truth from the false, rather than by guesswork or by something supernatural or from beyond the bounds of nature. It is this method of discovery, and the justification for that discovery, which must be accomplished with the utmost of integrity (5).

A look at how science and policy-making interact on the issue of global climate change illustrates both the limits to science and how science can, and should, influence action. Science is a body of knowledge, observations about the physical environment. Science is a systematic description of knowledge—theories that attempt to organize information into rules or laws. Science is also a modeling tool that helps us stretch time to look into the future and predict outcomes. And, perhaps most important, science is a set of processes that comprise a paradigm for investigation of the environment—a method of study.

Science is only one among many ways of studying the world around us. If we forget that, and expect science to give us an absolute answer to a problem, we invite misunderstanding and consequent policy errors. When we demand certainty from science before setting policy or acting on problems, we ask too much of science and expect too little from policy makers. As a body of knowledge, science is limited by our ability to observe and measure the physical world. It is also limited by the sheer volume of data available to us, and the inconsistent quality of data. It is expensive to collect, verify, process, and store data. The act of measurement and presence of the observer change the thing being investigated, sometimes destroying it. Consequently, we are often forced to make inferences about the whole based on partial knowledge.

To explain our observations, we develop theories. Scientific theories provide a framework for identifying critical uncertainties and focusing our quest for new information. A proposed theory is rejected if it fails to explain the facts at hand. Science thus advances by gathering new information and testing the existing body of science against it. When a scientific theory can no longer explain all of the data, or be reproduced by other competent analysts, it is rejected in favor of one that better explains the data. Science progresses by error, deduction and revision, through a process of debate and argument.

The branch of science that gets the most attention of late is that branch which makes models of reality and predictions about the future. When applied to climate science these models employ *Newtonian physics* and the vagaries of global weather thermal dynamics. These models are contentious, simply because they rely so heavily on subjective judgments about what features of a highly complex system to include in a model, how to interpret theories, and what data to consider. Because modeling, such as climate modeling, is subjective, it is controversial, and therefore, newsworthy.

On complex issues like global climate change, what typically gets reported is not the consensus among scientists about major underlying assumptions, the main body of knowledge, the solid theories that seem to be robust, or the best models. The disputes make better sound bites and more interesting headlines. The resulting appearance of chaos is inaccurate. It is important that we keep the differences in mind so that we don't fall into the trap of wrongly believing that science has final, absolute answers. Simply put, science cannot provide the sort of truth that can help us make error-free policy, nor should we expect it to. We certainly don't demand that policies related to non-technical matters be error-free. We don't require absolute certainty from economists before setting economic policy, nor from educators before designing education policy.

When we design policies in the face of uncertainty—i.e., virtually *all* policies—we should design them to permit monitoring outcome and results, so that we can learn from our actions. Policies should be reversible, in case we have missed a critical uncertainty, to reduce the consequences of being wrong. In environmental circles, this is called "adaptive management." (Other people have called it common sense.) Interestingly, when we design actions that are consistent with the adaptive management criteria, they often make the most sense whether the underlying science is right or wrong.

As to the specifics of the global climate change, we have good data over a period of decades that demonstrates increasing regional concentrations of the greenhouse gas carbon dioxide in the atmosphere. But the information we have on the Earth's temperature is not as good, so we cannot yet demonstrate with the certainty that the global climate is either warming or cooling.

Scientific theory explains the observed increase in carbon dioxide and other greenhouse gases in the atmosphere and their observed effect on heat reflection and radiation. The theory says that the more greenhouse gases we add to the atmosphere, the more incoming solar radiation will be trapped, tending to raise the temperature of the Earth. The theory cannot be considered "absolutely correct." But it does explain most of what we now know about atmospheric carbon dioxide and its effect on heat, and no competing theory offers a better explanation at this time.

The best global climate models predict that the greenhouse gases already in the atmosphere, and those we are likely to add to it over time, will lead to an increase in the Earth's temperature and changes in precipitation patterns. The models cannot tell us much about regional changes in storm patterns, or predict precisely how much or how soon the climate will warm up. But, they do point toward a significant warming over the next decades. That is the limit of what science can now tell us.

Last year 60 scientists sent a letter to Prime Minister Stephen Harper of Canada, urging him to undertake "a proper assessment of recent developments in climate science" and disputing the contention that "a climate catastrophe is looming and humanity is the cause." The letter cautioned that "observational evidence does not support today's computer climate models" and warned that since the study of climate change is relatively new, "it may be many years yet before we properly understand the Earth's climate system."

Among those signing the letter were the former director of the US Weather Satellite Service, a hydrogeology and paleoclimatology specialist, the former director of research at the Royal Netherlands Meteorological Institute, a physicist of the Institute for Advanced Studies in Princeton, a senior scientist in climate studies at NASA's Marshall Space Flight Center, plus 55 other specialists in climate science and related disciplines. One NASA scientist stated that while the general trend of global warming exists, that doesn't make it "a problem we must wrestle with." To insist that any change in climate must be bad news "is to assume that the ... Earth's climate today is the optimal climate, the best climate that we could have." The planet's temperature has been fluctuating for millennia, he added. "I don't think it's within the power of human beings to assure that the climate does not change."

In 2003 there was a survey of 530 environmental scientists in 27 countries on topics related to global warming. One question asked: "To what extent do you agree or disagree that climate change is mostly the result of anthropogenic (human) causes?" On a scale of 1 (strongly agree) to 7 (strongly disagree), the average score was 3.62, reflecting no clear consensus.

Asked whether abrupt climate changes will wreak devastation in some areas of the world, the percentage of scientists strongly agreeing (9.1) was nearly identical to the percentage strongly disagreeing (9.0). Another question asked: To what degree might global warming prove beneficial for some societies? A striking 34 percent of the scientists answered 1 or 2 (a great degree of benefit); just 8.3 percent answered 6 or 7 (very little/no benefit). Plainly, the science isn't settled at this time.

Take the latest report of the UN Intergovernmental Panel on Climate Change (IPCC). Unlike its previous report in 2001, which foresaw a possible rise in sea levels over the next century of around 3 feet, the new report cuts that figure in half, to about 17 inches. Why the revision? "Mainly because of improved information," the IPCC notes in the fine print. It goes on to note that even its latest estimate involves some guesswork: "Understanding of these effects is too limited to assess their likelihood." The science is getting better, but it's far from settled. Or take the discovery this month that 1934, not 1998, was the hottest year in the continental US since record-keeping began. NASA's Space Studies quietly changed its ranking after a Canadian statistician discovered an error in the official calculations. Under the new data, five of the 10 hottest US years on record occurred before 1940; three were in the past decade.

Climate scientists are still trying to get the basics right. The latest issue of *Science* magazine notes that many researchers are only beginning to factor the planet's natural climate variations into their calculations. "Until now," reports *Science*, "climate forecasters who worry about what greenhouse gases could be doing to climate have ignored what's happening naturally…. In this issue, researchers take their first stab at forecasting climate a decade ahead with current conditions in mind." Their first stab, please note, not their last. The science of climate change is still young and unsettled. Years of trial and error are still to come. Al Gore notwithstanding, the debate is hardly over (30).

It makes very good sense, as policy, to implement cost-effective energy conservation and conversion measures to reduce harmful air pollutants, including greenhouse gases. Such action is in line with "adaptive management" principles: it is potentially reversible, we can monitor the results, and it makes sense (because it would save money, improve air quality, and reduce dependence on imported oil) even if the predicted climatic warming is wrong (6). We should also continue to fund and study climate science, gather more information about the Earth's temperature, test our theories against new information, and refine our climate models.

Climate Action Alternatives

There is no shortage of irrational, extremist, partisan and down right silly proposed action solutions from "experts" in the issues of global climate change. Bad environmental policy initiatives will always flow from bad science, especially in this period of evolving climate science and political opportunism. The following respected experts in the area of global climate change represent the three alternative actions in response, if any, to global warming.

Alternative 1—No Action, More Science

Richard S. Lindzen is the Alfred P. Sloan Professor of Meteorology at the Massachusetts Institute of Technology. Lindzen concedes that the world today regards "global warming" as both real and dangerous. Indeed, the diplomatic activity concerning warming might lead one to believe that it is the major crisis confronting mankind. The June 1992 Earth Summit in Rio de Janeiro, Brazil, focused on international agreements to deal with that threat, and the heads of state from dozens of countries attended. Lindzen can find no substantive basis for the warming scenarios being popularly described. Moreover, according to many studies by economists, agronomists, and hydrologists, there would be little difficulty adapting to such warming if it were to occur. Such was also the conclusion of the recent National Research Council's report on adapting to global change. Many aspects of the catastrophic scenario have already been largely discounted by the scientific community. For example, fears of massive sea-level increases accompanied many of the early discussions of global warming, but those estimates have been steadily reduced by orders of magnitude, and now it is widely agreed that even the potential contribution of warming to sea-level rise would be swamped by other more important factors. Lindzen asserts that there is no substantive basis for predictions of sizeable global warming due to observed increases in minor greenhouse gases such as carbon dioxide, methane, and chlorofluorocarbons, and that there is no basis for corrective action or for prediction of the results of any action (14). Lindzen summarizes: First, nonscientists generally do not want to bother with understanding the science. Claims of consensus relieve policy types, environmental advocates and politicians of any need to evolve science. Such claims also serve to intimidate the public and even scientists, especially those outside the area of climate dynamics. Secondly, given that the question of human causation largely cannot be resolved, their use in promoting visions of disaster constitutes nothing so much as a scam. Here is an inauspicious beginning to what Al Gore claims is not a political issue but a "moral" crusade. Lastly, there is a clear attempt to establish truth not by scientific methods but by perpetual repetition. An earlier attempt at this was accompanied by tragedy. Perhaps Marx was right. This time around we may have a farce—if we're lucky (11).

Bjorn Lomborg, Danish statistician, economist, former Greenpeace official, and author of the bestseller book "The Skeptical Environmentalist," says that global warming would have some good impacts. Because many more people die in the cold winters in areas above and below the equator. If the globe heats due to human activities, then fewer people will die in winter. Lomborg offers numerous other potentially positive impacts of global warming in his book. He fully believes that man contributes to global warming, but that the media attempts to stampede us into what may be irrational and very costly reactions is folly. We'er talking about a temperature rise of less than a quarter of a degree over the next five years. Lomborg urges not to challenge the reality of global warming or the fact that it's caused in large part by humans. He says that the discussion about climate change has turned into a nasty dustup, with one side arguing that we're headed for catastrophe and the other maintaining that it's all a hoax—he says that neither is right. It's wrong to deny the obvious: The Earth is warming, and we're causing it. But that's not the whole story, and predictions of impending disaster just don't stack up. We have to rediscover the middle ground, where we can have a sensible conversation. We shouldn't ignore climate change or the policies that could attack it. But we should be honest about the shortcomings and costs of those policies, as well as the benefits (12).

Patrick J. Michaels is a Senior Fellow in environmental studies at the Cato Institute and author of the book "Meltdown: The Predictable Distortion of Global Warming by Scientists, Politicians and Media." Michaels says that even if every nation on Earth achieved the Kyoto Protocols greenhouse gas controls, it would prevent no more than 0.126 degrees F of warming over 50 years. Global temperatures vary by more than this from year to year. The reality is that if we really want to control the process of global warming significantly, we have to cut greenhouse gas emissions by an extremely large amount—perhaps by 40% or more. However, such control technology does not now exist. So, should we fritter away billions in precious investment capital in a futile attempt to control warming? Michaels suggests that the best course of action is to live with some modest climate change now and encourage economic development, which would generate the capital necessary for investment in the cost-effective control technologies of the future (43).

Alternative 2—Act Now on Feasible Reductions

Most agree that accumulating carbon dioxide and other heat-trapping smokestack and tailpipe gases pose a significant environmental challenge, but they also agree that the appropriate response is more akin to buying fire insurance, installing sprinklers and new wiring in an old, irreplaceable house (the Earth) than to fighting a fire already raging.

"Climate change presents a very real risk," said Carl Wunsch, a climate and oceans expert at the Massachusetts Institute of Technology. "It seems worth a very large premium to insure ourselves against the most catastrophic scenarios. Denying the risk seems utterly stupid. But, claiming that we can calculate the probabilities of these scenarios with any degree of skill seems equally stupid."

Many in this camp seek a policy of reducing vulnerability to all climate extremes while building public support for a sustained shift to nonpolluting energy sources. They have made their voices heard in Web logs, news media interviews and at least one statement from a large scientific group, the World Meteorological Organization. In early December 2006, this group posted a statement written by a committee consisting of most of the climatologists assessing whether warming seas have affected hurricanes. While each degree of warming of tropical oceans is likely to intensify such storms a percentage point or two in the future, they said, there is no firm evidence of a heat-triggered strengthening in storms in recent years. The experts added that the recent increase in the impact of storms was because of more people getting in harm's way, not stronger storms.

There are enough experts holding such views that Roger A. Pielke Jr., a political scientist and blogger at the University of Colorado, Boulder, came up with a name for them (and himself): "nonskeptical heretics." "A lot of people have independently come to the same sort of conclusion," Dr. Pielke said. "We do have a problem, we do need to act, but what actions are practical and pragmatic?"

This approach was most publicly laid out in an opinion article on the BBC Website in November by Mike Hulme, the director of the Tyndall Center for Climate Change Research in the UK. Dr. Hulme said that shrill voices crying doom could paralyze instead of inspire.

"I have found myself increasingly chastised by climate change campaigners when my public statements and lectures on climate change have not satis- fied their thirst for environmental drama," he wrote. "I believe climate change is real, must be faced and action taken. But the discourse of catas- trophe is in danger of tipping society onto a negative, depressive and reac- tionary trajectory." Other experts say there is no time for nuance, given the general lack of public response to the threat posed particularly by carbon dioxide, a byproduct of burning fossil fuels and forests that persists for a century or more in the air and is accumulating rapidly in the atmosphere.

Debate among scientists over how to describe the climate threat is particularly intense right now as experts work on the final language in portions of the latest assessment of global warming by the UN Intergovernmental Panel on Climate Change (see Chapter 6 herein). In three previous reports, this global network of scientists has presented an ever-firmer picture of a growing human role in warming. Studies used to generate the next report have shown a likely warming in the 21st century—unless emissions of greenhouse gases abate—at least several times that of the last century's one-degree rise. But substantial uncertainty still clouds projections of important impacts, like how high and quickly seas would rise as ice sheets thawed.

Other drafts of the climate report used a conservative analysis that does not project a rise most people would equate with catastrophe. Other experts say this may send too comforting a message. Dr. Hulme insists that it is best not to gloss over uncertainties. In fact, he and other experts say that uncertainty is one reason to act as a hedge against the prospect that problems could be much worse than projected. His goal, Dr. Hulme said, is to raise public appreciation of the unprecedented scale and nature of the challenge. "Climate change is not a problem waiting for a solution (least of all a solution delivered and packaged by science), but a powerful idea that will transform the way we develop," he said. Dr. Hulme and others avoid sounding alarmist, but offer scant comfort to anyone who doubts that humans are contributing to warming or believes the matter can be deferred. These experts see a clear need for the public to engage now, but not to panic. They worry that portrayals of the issue like that in "An Inconvenient Truth," the documentary focused on the views of Al Gore, may push too hard. Many in this group also see a need to portray clearly that the response would require far more than switching to fluorescent light bulbs and to hybrid cars.

"This is a mega-ethical challenge," said Jerry D. Mahlman, a climatologist at the National Center for Atmospheric Research in Boulder, Colo., who has studied global warming for more than three decades. "In space, it's the size of a planet, and in time, it has scales far broader than what we mere *Homo sapiens* are accustomed to dealing with." Dr. Mahlman and others say that the buildup of carbon dioxide and other greenhouse gases cannot be quickly reversed with existing technologies. And even if every engine on Earth were shut down today, they add, there would be no measurable impact on the warming rate for many years, given the buildup of heat already banked in the seas. Because of the scale and time lag, a better strategy, Dr. Mahlman and others say, is to treat human-caused warming more as a risk to be reduced than a problem to be solved (10).

Alternative 3—Act Now, Science and Costs Later

World renowned Indian environmentalist and UN Intergovernmental Panel on Climate Change (IPCC) Chairman R. K. Pachauri has warned international business and political leaders that climate change could cost up to 5 per cent of global GDP by 2030 if effective steps were not taken in time. Addressing the annual meeting of World Economic Forum, Pachauri said, the threat of climate change has become a threat to world peace. Though there is an unprecedented awareness about the issue, the response of corporate sector and governments was still weak.

Expressing disappointment at global response to the challenge, he said the new EU strategy for slashing greenhouse emissions was "not up to expectations." The new IPCC chairman also said, the use of renewable energies like biomass, wind and solar power should rise to 20 per cent of all energy forms over the next two decades. Biofuels would also have to make up 10 per cent of fuels used for transport.

Pachauri, whose organization won 2007's *Nobel Peace Prize* with former US vice president Al Gore, said he was sure that the priorities would change over a period of time. As per the estimates of IPCC, which has studied the impact of climate change on environment, he said, "The global economy incurred a loss of around 1.5 trillion dollars during 1982–2004 due to natural disasters like floods, drought and loss in crops due to climate change."

Pachauri said despite certain measures taken by the global community, the emission of gases like methane and carbon monoxide have not declined so far. The impact of global warming was already visible through rise in sea levels, environment change and heat waves. He said the business leaders should realize that they would also get economic benefits if they took measures to fight with emission of green gases. Otherwise, it would result in their loss of reputation, and produce a political backlash due to wide public awareness (15).

Al Gore has been the most visible and vocal proponent of radical action to stop global warming. And, he has profited both politically and financially from his climate crusade. One of his more fanciful and flamboyant eco-stunts was the 2007 "Live Earth" global music festival and fundraiser. Here, Gore insisted that we take immediate actions as itemized in the seven-step "pledge" below:

1. Demand that your country join an international treaty within the next two years that cuts global warming pollution by 90 percent in developed countries and by more than half worldwide in time for the next generation to inherit a healthy Earth.

2. Take personal action to help solve the climate crises by reducing your own CO2 pollution as much as you can, and by offsetting the rest to become "carbon neutral."

3. Fight for a moratorium on the construction of any new power generating facility that burns coal without the capacity to safely trap and store the CO2.

4. Work for a dramatic increase in the energy efficiency of your home, workplace, school, place of worship and means of transportation.

5. Fight for laws and policies that expand the use of renewable energy sources and reduce dependence on oil and coal.

6. Plant new trees and join with others in preserving and protecting forests.

7. Buy from businesses and support leaders who share your commitment to solving the climate crises and building a sustainable, just and prosperous world for the 21st century (16).

As with other shortsighted, feel-good "green solutions," Gore provides no mention of the personal or global costs or actual measurable benefits of the pledge actions.

CHAPTER 5

▼

CLIMATE SILLINESS

The issues of global warming and climate change have spawned a lot of hasty science, green product marketing, frantic government policy and half-baked climate change solutions. These well-intentioned solutions offer theories requiring business and personal lifestyle changes that are often bizarre, polluting, prohibitive in cost, and even silly.

The following over 600 items have been published as being caused by global warming:

> acne, agricultural land increase, Africa devastated, African aid threatened, Africa hit hardest, air pressure changes, Alaska reshaped, allergies increase, Alps melting, Amazon a desert, American dream end, amphibians breeding earlier (or not), ancient forests dramatically changed, animals head for the hills, Antarctic grass flourishes, Antarctic ice grows, Antarctic ice shrinks, anxiety, algal blooms, archaeological sites threatened, Arctic bogs melt, Arctic in bloom, Arctic ice free, Arctic lakes disappear, asthma, Atlantic less salty, Atlantic more salty, atmospheric defiance, atmospheric circulation modified, attack of the killer jellyfish, avalanches reduced, avalanches increased, Baghdad snow, bananas destroyed, bananas grow, beetle infestation, bet for

$10,000, better beer, big melt faster, billion dollar research projects, billions face risk, billions of deaths, bird distributions change, bird visitors drop, birds return early, birds driven north, blackbirds stop singing, blizzards, blue mussels return, bluetongue, boredom, bridge collapse (Minneapolis), Britain Siberian, British gardens change, brothels struggle, bubonic plague, budget increases, Buddhist temple threatened, building collapse, building season extension, bushfires, business opportunities, business risks, butterflies move north, camel deaths, cancer deaths in England, cardiac arrest, caterpillar biomass shift, cave paintings threatened, challenges and opportunities, childhood insomnia, Cholera, circumcision in decline, cirrus disappearance, civil unrest, cloud increase, cloud stripping, cockroach migration, cod go south, cold climate creatures survive, cold spells (Australia), computer models, conferences, coral bleaching, coral reefs dying, coral reefs grow, coral reefs shrink, cold spells, cost of trillions, cougar attacks, cremation to end, crime increase, crocodile sex, crumbling roads, buildings and sewage systems, cyclones (Australia), damages equivalent to $200 billion, Darfur, Dartford Warbler plague, death rate increase (US), Dengue hemorrhagic fever, dermatitis, desert advance, desert life threatened, desert retreat, destruction of the environment, diarrhoea, disappearance of coastal cities, diseases move north, Dolomites collapse, drought, drowning people, ducks and geese decline, dust bowl in the corn belt, early marriages, early spring, earlier pollen season, Earth biodiversity crisis, Earth dying, Earth even hotter, Earth light dimming, Earth lopsided, Earth melting, Earth morbid fever, Earth on fast track, Earth past point of no return, Earth slowing down, Earth spinning out of control, Earth spins faster, Earth to explode, Earth upside down, Earth wobbling, earthquakes, El Niño intensification, end of the world as we know it, erosion, emerging infections, encephalitis, equality threatened, Europe simultaneously baking and freezing, evolution accelerating, expansion of university climate groups, extinctions (human, civilisation, logic, Inuit, smallest butterfly, cod, ladybirds, bats, pandas, pikas, polar bears, pigmy possums, gorillas, koalas, walrus, whales, frogs, toads, turtles, orang-utan, elephants, tigers, plants, salmon, trout, wild flowers, woodlice, penguins, a million species, half of all animal and plant species, mountain species, not polar bears, barrier reef, leaches), experts muzzled, extreme changes to California, fading fall foliage, famine, farmers go under, fashion disaster, fever, figurehead sacked, fir cone bonanza, fish catches drop, fish downsize, fish catches rise, fish stocks at risk, fish stocks decline, five million illnesses, flesh eating disease, flood patterns change, floods, floods of beaches and cities, Florida economic decline, flowers in peril, food poisoning, food prices rise, food prices soar, food security threat (SA), footpath erosion, forest decline, forest expansion, frostbite, frosts, fungi fruitful, fungi invasion, games change, Garden of Eden wilts, genetic diversity decline, gene pools slashed, giant squid migrate, gingerbread houses collapse, glacial earthquakes, glacial retreat, glacial growth, glacier wrapped, global cooling, global dimming, glowing clouds, god melts, golf Masters wrecked, Gore omnipresence, grandstanding, grasslands wetter, Great Barrier Reef 95% dead, Great Lakes drop, greening of the North, Grey whales lose weight, Gulf Stream failure, habitat loss, Hantavirus pulmonary syndrome, harvest increase, harvest shrinkage, hay fever epidemic, hazardous waste sites breached, health of children harmed, heart disease, heart attacks and strokes (Australia), heat waves, hibernation ends too soon, hibernation ends too late, homeless 50 million, hornets, high court debates, human development faces unprecedented reversal, human fertility reduced, human health improvement, human health risk, human race oblivion, hurricanes, hurricane reduction, hydropower problems, hyperthermia deaths, ice sheet growth, ice sheet shrinkage, illness and death, inclement weather, infrastructure failure (Canada), Inuit displacement, Inuit poisoned, Inuit suing, industry threatened, infectious diseases, inflation in China, insurance premium rises, invasion of cats, invasion of herons, invasion of midges, island disappears, islands sinking, itchier poison ivy, jellyfish explosion, Kew Gardens taxed, kitten boom, krill decline, lake and stream productivity decline, lake shrinking and growing, landslides, landslides of ice at 140 mph, lawsuits increase, lawsuit successful, lawyers' income increased (surprise surprise!), lightning related insurance claims, little response in the atmosphere, lush growth in rain forests, Lyme disease, Malaria, malnutrition, mammoth dung melt, Maple syrup shortage, marine diseases, marine food chain decimated, marine dead zone, Meaching (end of the world), megacryometeors, Melanoma, methane emissions from plants, methane burps, melting permafrost, Middle Kingdom convulses, migration, migration difficult (birds), microbes to decompose soil carbon more rapidly, monkeys on the move, Mont Blanc grows, monuments imperiled, more bad air days, more research needed, mortality increased, mountain (Everest) shrinking, mountains break up, mountains taller, mortality lower, mudslides, National security implications, natural disasters quadruple, new islands, next ice age, Nile delta damaged, noctilucent clouds, no effect in India, Northwest Passage opened, nuclear plants bloom, oaks dying, oaks move north, oblivion, ocean acidification, ocean waves speed up, opera house to be destroyed, outdoor hockey threatened, oyster diseases, ozone loss, ozone repair slowed, ozone rise, Pacific dead zone, personal carbon rationing, pest outbreaks, pests increase, phenology shifts, plankton blooms, plankton destabilised, plankton loss, plant viruses, plants march north, polar bears aggressive, polar bears cannibalistic, polar bears drowning, polar bears starve, polar tours scrapped, porpoise astray, profits collapse, psychosocial disturbances, puffin decline, railroad tracks deformed, rainfall increase, rainfall reduction, rape wave, refugees, reindeer larger, release of ancient frozen viruses, resorts disappear, rice threatened, rice yields crash, riches, rift on Capitol Hill, rioting and nuclear war, rivers dry up, river flow impacted, rivers raised, roads wear out, rockfalls, rocky peaks crack apart, roof of the world a desert, rooftop bars, Ross river disease, ruins ruined, salinity reduction, salinity increase, Salmonella, salmon stronger, satellites accelerate, school closures, sea level rise, sea level rise faster, seals mating more, sewer bills rise, severe thunderstorms, sex change, sharks booming, sharks moving north, sheep shrink, shop closures, shrimp sex problems, shrinking ponds, shrinking shrine, ski resorts threatened, slow death, smaller brains, smog, snowfall increase, snowfall heavy, snowfall reduction, soaring food prices, societal collapse, song-birds change eating habits, sour grapes, space problem, spiders invade Scotland, squid population explosion, squirrels reproduce earlier, spectacular orchids, storms wetter, stormwater drains stressed, street crime to increase, suicide, Tabasco tragedy, taxes, tectonic plate movement, teenage drinking, terrorism, threat to peace, ticks move northward (Sweden), tides rise, tourism increase, trade barriers, trade winds weakened, tree beetle attacks, tree foliage increase (UK), tree growth slowed, trees could return to Antarctic, trees in trouble, trees less colourful, trees more colourful, trees lush, tropics expansion, tropopause raised, tsunamis, turtles crash, turtles lay earlier, UK coastal impact, UK Katrina, Vampire moths, Venice flooded, volcanic eruptions, walrus displaced, walrus pups orphaned, walrus stampede, war, wars over water, wars sparked, wars threaten billions, water bills double, water supply unreliability, water scarcity (20% of increase), water stress, weather out of its mind, weather patterns awry, weeds, Western aid cancelled out, West Nile fever, whales move north, wheat yields crushed in Australia, white Christmas dream ends, wildfires, wind shift, wind reduced, wine—harm to Australian industry, wine industry damage (California), wine industry disaster (US), wine—more English, wine—German boon, wine—no more French , winters in Britain colder, wolves eat more moose, wolves eat less, workers laid off, World bankruptcy, World in crisis, World in flames, Yellow fever (2).

Some of the popular solutions to the above-listed global warming maladies involve renewable energy sources, "carbon footprinting," cow flatulence control, changing of light bulbs, and other fanciful remedies. These solutions, or mitigation measures, focus on general energy reduction, and specifically, reductions in greenhouse gases. Most of these measures fail the basic tests of economic scale, actual greenhouse gas reductions and practical enforcement. The following items are examples of the silly, to the bizarre, solutions for global warming problems:

- Renewable energy sources have become much the rage of world governments and eco-activists. These energy technologies or sources include biofuels, hydropower, geothermal, wind and solar. Recent researchers conclude that renewables aren't green because to reach the scale at which they would contribute importantly to meeting global energy demand, renewable sources of energy such as biofuels, hydropower, wind and solar, cause serious environmental harm. For example, it would take a wind farm more than 475 square miles in size to generate the same amount of electrical power that a single, shopping center-sized 1,000-megawatt nuclear plant would produce. Put another way, an area the size of Texas would have to be covered with windmills running 24 hours a day, along with the structures needed to store and transmit the energy, to generate enough power to meet the 2005 US electricity demand. Biomass is even worse—chewing up three to 10 times as much space as wind power.

  Hydropower would take every drop of the yearly rainfall in Ontario, Canada—180,000 billion gallons—sitting behind a dam 200 feet high to provide 80% of the power supplied by that country's 25 nuclear power plants. As for solar, the entire state of Connecticut would have to be covered in solar cells and associated retrieval and transmission structures just to provide power for New York City (20).

- Almost all biofuels, including ethanol, cause more greenhouse gas emissions than conventional fuels if the full emissions costs of producing theses "green" fuels are taken into account. In the US, politicians who whine about the price of gasoline fall over themselves to support ethanol, which drives up the prices of farmed foods . We now subsidize ethanol production to the tune of $7 billion a year. This encourages crop production for fuel, not food, and the competition for a finite crop of corn drives up the price of everything at the supermarket.

It also comes at an environmental cost. Corn farmland rainfall run-off due to the rush to increase corn production has contributed to the creation of dead zones in the Gulf of Mexico and elsewhere along our coasts. An oxygen-poor dead zone, created by oxygen-sucking algae fed by fertilizers used to grow corn and other bio-fuels in the Midwest watershed of the Mississippi River already exists in the Gulf of Mexico. The 7,900-square-mile area with almost no oxygen—a condition called hypoxia—is about the size of Connecticut and Delaware together.

Ethanol from corn sounds like an energy panacea, but the devil is in the details. It takes 4,000 gallons of fresh water per acre per day to replace evaporation in a cornfield. Each acre requires about 130 pounds of nitrogen and 55 pounds of phosphorous fertilizers. Never mind the fossil fuel used by farm equipment needed to plow, cultivate and harvest the crop. Then the corn must be refined into ethanol that produces 20% to 30% less energy than gasoline, and it must be transported around the country by truck, not pipelined. In the rush to develop these alternative fuels, forests in Asia have been burned to clear land for palm oil, and large swaths of the Amazon rain forest are being stripped of diverse vegetation for soy and sugar plantations used to produce the raw material for making ethanol (18).

- Consumer choices and budgets are coming under *green government* scrutiny. Several British food companies and retailers plan to add "carbon footprint" labels showing the quantity of carbon dioxide emissions associated with making and transporting foods and other goods. These labels were produced in conjunction with the Carbon Trust, an environmental consultancy funded by the British government. The consumers green guilt would be tested by the carbon footprint of the product purchased. The labels do not count the energy needed for refrigeration, lighting and heating in the retail stores. Nor do they include the greenhouse gas emissions that come from using a product. A particularly difficult area is the new and obscure field of agricultural modeling. A recent report funded by DEFRA, Britain's environment agency, found that some organic foods had larger carbon footprints than conventional ones. It was criticized by the Soil Association, Britain's main organic foods lobby, which took issue with the models used for the calculations (21).

- The UN has gone to great lengths to identify even the most obscure greenhouse gas sources that threaten global survival. Titled "Livestock's Long Shadow," a UN's Food and Agriculture Organization report assesses the damage done by sheep, chickens, pigs and goats. But it mostly puts the blame on the world's 1.5 billion cows. Altogether, the report says, flatulent livestock and farming are the source of 18% of the greenhouse gases—more than cars, planes and all other forms of transport put together. The report has created quite a stink by documenting the impact on Earth's climate by gases from manure and flatulence, deforestation (including destruction of rain forests to create grazing land) and the energy used in farming. North America alone has more than 100 million cattle, hundreds of millions of hogs and pigs, and more than 2 billion chickens, all emitting billions of tons of greenhouses gases each year. The UN said cows and other critters produce methane—lots of it—and each molecule of methane has 21 times as much global warming impact as a molecule of carbon dioxide (25).

- The National Association of Evangelicals has gone a long way towards transforming the thinking of American Christians about the planet and environmentalism. But, *green evangelicals* have faced an uphill battle. Their preference for the term "creation care" is a reflection, in part, of the horror that the word "environmentalism"—tinged with secular humanism, even liberal paganism—still strikes at the evangelical psyche. But, according to a poll in 2007, two-thirds of evangelicals want immediate action on global warming. Creation care is only part of it: the urge to enhance national security by lessening America's dependence on foreign oil is important too. One open question is how far green evangelicals can go in co-operating with Christians with a different ethos. But some intriguing exchanges may be triggered by the recent conversion of Pope Benedict XVI—widely admired by American conservatives—to the view that global warming is a serious issue. Already, even within the evangelical camp, some patching over of differences has been needed to rally round the green cause: some eco-evangelicals root their beliefs in a literal reading of the book of Genesis, while others interpret the Creation story a bit more abstractly (28).

- Enviro-fanatics are sterilizing themselves to reduce their "carbon footprint." This from a British environmental activist who had sterilization surgery at the age of 27 because she considers children "a sinister threat to the future." Two years earlier, despite being on birth-control pills, the eco-freak got pregnant and had an abortion because, "… it would have been immoral to give birth to a child that I felt strongly would only be a burden to the world." She further proclaimed "having children is selfish." A vegetarian by age 15, she met her husband at an animal rights rally. On the morning of her sterilization, her husband gave her a "Congratulations" card. Each new child, she says, "uses more food, more water, more land, more fossil fuels, more trees and produces more rubbish, more pollution, more greenhouse gases, and adds to the problem of over-population." What a depressing take on the value of human life. But it was echoed by another eco-crazed couple. Sarah Irving, a 31-year-old *green magazine* editor, and Mark Hudson, a 37-year-old health care worker, reminisced about how, "after a year of dating, we started talking about sterilization." According to Hudson, they "lived as green a life as possible"—no car, low-energy light bulbs and only locally grown organic foods. They "cycle everywhere" and "never fly." In short, he says, "we do everything we can to reduce our carbon footprint. But all this would be undone if we had a child. That's why I had a vasectomy. It would be morally wrong for me to add to climate change and the destruction of Earth.... What makes us happy is knowing that we are doing our bit to save our precious planet" (29).

CHAPTER 6

▼

FORWARD CLIMATE POLICY

The most important future global environmental policies should promote the sharing of cost-effective pollution control technologies, and the expansion of 21st Century electric power generation (including nuclear power) into underdeveloped and developing countries. The US Government, in partnership with commercial technology companies, should adopt and fully fund for the next electric power generation systems with the resolve, vigor and national pride of the *Manhattan Project* that produced the first nuclear bomb in the 1940s.

The US and other industrialized countries are still using 19th Century electric power generation technologies in the 21st Century. In the US, only 4% of the nation's electricity is generated by oil, compared with 52% by coal, 15% by natural gas, 19% by nuclear reactors and less than 10% by renewable energy sources such as wind, solar, hydroelectric, etc. (41). Today, 103 nuclear power plants account for 19% of the US electric power supply. One half of the US uranium that powers those plants comes from recycled Russian nuclear weapons *Cold War* disarmaments. There are 430 nuclear power plants operating in 31 countries worldwide. The energy from one pound of uranium is equivalent to 1.3 million pounds of coal energy. Nuclear power produces none of the greenhouse gases associated with global warming.

The 1986 nuclear reactor accident at Chernobyl in Ukraine spread radioactivity over Europe and despair in the Western world's nuclear industry. However, some countries never lost their enthusiasm for nuclear power. Nuclear provides 80% of French electricity, and some developing countries have continued to build nuclear plants. But elsewhere in the West, Chernobyl, along with the accident at Three Mile Island in Pennsylvania in 1979, sent the industry into a decline. The public got scared. The regulatory environment tightened, raising nuclear power costs. Billions were spent bailing out nuclear power companies. The industry became a byword for mendacity, secrecy and profligacy with taxpayer money. For two decades neither governments nor bankers wanted to touch it. Now nuclear power has a second chance. Its revival is most visible in the US, where power companies are preparing to flood the Nuclear Regulatory Commission with applications to build new plants. And in other countries, Finland is building a reactor, the UK is preparing the way for new nuclear planning regulations. In Australia, which has plenty of uranium but no reactors, nuclear power is deemed inevitable.

Geopolitics, technology, economics and environmentalists are moving toward nuclear power. Western governments are concerned that most of the world's oil and gas is in the hands of hostile or unreliable governments. Much of the nuclear industry's raw material, uranium, is conveniently located in friendly places such as Australia and Canada. Simpler reactor designs cut maintenance and repair costs. Shut-downs are now far less frequent, so that a typical US nuclear power plant operates 90% of the time, up from less than 50% in the 1970s. New "passive safety" features can shut a reactor down in an emergency without the need for human intervention. The US plans to embrace a new approach in which the most radioactive portion of the reactor waste from conventional nuclear power plants is isolated and burned in "fast" reactors. Technology has thus improved nuclear power's economics, as has the squeeze of fossil fuel price escalation. Nuclear power stations are expensive to build but very cheap to run. Conventional oil-bases power plants—the bulk of new build in the 1980s and 1990s—are the reverse. Since oil fuels provide the extra power needed when demand rises, the oil price sets the electricity price. Proposed *carbon taxes* or *carbon trading* (presently estimated at $20 to $40 per ton) would significantly inflate the production costs of conventional coal and oil-based power plants, and make nuclear more viable. Costly oil has therefore made nuclear power not only competitive, but quite profitable. Any pricing of carbon emissions will cause a global re-structuring of public power facilities and a commensurate re-pricing in global energy markets.

The hyper interest in climate change has also made nuclear power attractive. Nuclear power offers the possibility of large quantities of baseload electricity that is cleaner than coal, more secure than gas and more reliable than wind or solar energy alternatives. And if cars switch from oil-based fuels to electricity, the demand for power generated from carbon-free sources (including nuclear) will increase still further. The nuclear industry's image is thus turning from black to green. A recent UK poll showed 30% of the population against nuclear power, compared with 60% three years ago. A US poll in 2007 showed 50% in favor of expanding nuclear power, up from 44% in 2001 (32).

The major advances in global policy initiatives to address global climate change have come from the United Nations (UN), the European Union (EU) and the United States (US). The following section presents summaries of the policy initiatives, and their estimated costs of implementation, for global climate change actions of those governing bodies.

The United Nations

Climate change is a very complex issue: policymakers need information about the causes of climate change, its potential environmental and socio-economic consequences and the adaptation and mitigation options to respond to it. This is why the United Nations (UN) established the

Intergovernmental Panel on Climate Change (IPCC) in 1988. The IPCC is a scientific body: the information it provides with its reports is based on scientific evidence and reflects existing viewpoints within the scientific community. The comprehensiveness of the scientific content is achieved through contributions from experts in all regions of the world and all relevant disciplines including, where appropriately documented, industry literature and traditional practices, and a two stage review process by experts and governments. When governments accept the IPCC reports and approve their "Summary for Policymakers," they acknowledge the legitimacy of their scientific content.

The IPCC provides its reports at regular intervals and they immediately become standard works of reference, widely used by policymakers, experts and students. The findings of the First IPCC Assessment Report of 1990 played a decisive role in leading to the UN's Framework Convention on Climate Change (UNFCCC), which was opened for signature in the Rio de Janeiro Summit in 1992 and entered into force in 1994. It provides the overall policy framework for addressing the climate change issue. The IPCC Second Assessment Report of 1995 provided key input for the negotiations of the Kyoto Protocols in 1997 and the Third Assessment Report of 2001 as well as Special and Methodology Reports provided further information relevant for the development of the UNFCCC and the Kyoto Protocols (35). The Kyoto Protocols expire in 2012.

The IPCC Forth Annual Assessment Report was released in full in May 2007—it aims to estimate the costs and benefits of various climate change policies. The report suggests that if major climate impacts are to be avoided, global emissions should peak and begin declining within one or two decades. Greenhouse gas emissions have risen by 70% since 1970, and will rise by between 25% and 90% over the next 25 years under "business as usual." That rise will mainly be caused by an expansion in the use of fossil fuels, which are set to continue as the world's dominant energy source.

The report assesses the likely costs to the global economy of stabilizing greenhouse gases at various concentrations in the atmosphere. The sharpest cuts, keeping greenhouse gas concentrations to levels equivalent to between 445 and 535 parts per million of carbon dioxide (CO_2), might cost as much as 3% of global GDP by 2030. The current atmospheric concentrations are equivalent to about 425 parts per million of CO_2. Assessing the impacts of a given concentration is not an exact science, but many researchers believe that keeping concentrations below about 450ppm CO_2 is necessary if the average global temperature rise is to be kept below 2C (3.6F), and major impacts avoided. The IPCC suggests that concentrations between 445ppm and 490ppm would keep the temperature rise to 2.0–2.8C (3.6–5.0F). European Union policy is to avoid a temperature rise greater than 2C (3.6F) (1).

The European Union

The European Union (EU) revealed its climate change policy in January 2008. The massive climate change plan details how much pain each of its 27 member countries will have to bear if the EU is to meet ambitious targets set by national leaders in 2007. The aim is to cut greenhouse gas emissions by 2020 by at least a fifth, and to more than double to 20% the amount of energy produced from renewable sources such as solar, wind or hydropower. If biofuel from plants proves green enough, 10% of the fuel used in transportation must come from biofuels by the 2020 target. The new plan turns these goals into national targets. The European Commission's recommendations must now be adopted into binding law by national governments and the European Parliament. Wealthier EU countries will be asked to take more of the burden than newer members. Sweden, for example, would have to meet 49% of its energy from renewables. While, Malta gets a renewables target of just 10%. It is a similar story when it comes to cutting greenhouse gases. By 2020, Denmark must cut emissions by 20% from 2005 levels; Bulgaria and Romania, the newest EU members, may let their emissions rise by 20%.

The new EU initiatives on climate change will not come cheap. The direct costs alone are estimated at €60 billion ($87 billion), or about 0.5% of total EU GDP, by 2020. The EU approach relies heavily upon its emissions trading scheme (ETS), which, in its several years of operation, has not met market or environmental performance objectives. The ETS obliges big polluters such as power companies or manufacturing plants to buy and trade permits that allow them to emit CO2 and other greenhouse gases, within a steadily tightening overall cap. If countries such as the US and China do not sign up to binding international agreements by 2011, then the heaviest greenhouse gas emitters inside the EU may be given these emission permits at no cost (13).

The United States

The US has been far more cautious than its fellow UN and EU climate crusaders. This is due to the unknowns in nascent climate science, and the real fears of the negative economic impacts from additional climate regulations on top of the US's current environmental control costs of about 5% of GDP. And fears that the developing economic giants of China and India, that now produce more greenhouse gases than the US or any other country, will not accept meaningful climate change regulations. The 1997 Kyoto Protocols, which has been accepted by many nations, have not been accepted by the US, China or India. Available estimates of the costs of US compliance with "Kyoto-like" greenhouse gas controls are at 2% of GDP—$260 billion each year, totaling more than $11 trillion by 2050 (26).

A new US climate change initiative was adopted in late 2007 as the Energy Independence and National Security Act of 2007. The Act would establish renewable fuels standards and automotive fuel efficient measures to mitigate greenhouse gases and dependency upon foreign oil. Some of the key provisions of this Act include:

- Reduction in US oil use of 2.8 million barrels a day by 2020, and 5 million barrels per day by 2030, over business as usual;

- US consumer fuel savings of $71 billion per year in 2020, and $161 billion in 2030, using approximate current prices ($90/barrel oil, $3/gallon gasoline);

- Reduction of transfer of wealth abroad of $73 billion per year in 2020 and $129 billion in 2030, using the same prices;

- Reduction in US CO2 emissions by 320 million metric tons in 2020, and 675 million metric tons in 2030, representing a reduction in passenger vehicle emissions of 15% and 30%, respectively, from what they otherwise would be;

- Reduction in 2020 of approximately 4% of projected total net US CO2 emissions versus what they would otherwise be (22).

Several US legislative initiatives to impose direct greenhouse gas controls by way of a "carbon tax" or a "cap and trade" program as adopted by other developed countries have failed to pass. And, the US Environmental Protection Agency (EPA) has been sued by eco-groups to designate carbon dioxide as an "air pollutant" under the almost forty year old federal *Clean Air Act.*

The US climate change policy reads as follows:
"The United States Federal government has established a comprehensive policy to address climate change. This policy has three basic objectives:

- Slowing the growth of emissions;

- Strengthening science, technology and institutions;

- Enhancing international cooperation.

The Federal government is implementing this policy through voluntary and incentive-based programs and has established major government-wide programs to advance climate technologies and improve climate science" (27).

The environmental movement and regulatory system reached its full, practical expression in the 20th Century. The movement and its environmentalists must concede that virtually every thing "regulatable" is regulated, and that the energy of environmental activism should 1) be directed globally to underdeveloped and developing countries, or 2) be directed to improvements in US security, education, electoral equity and welfare issues with moderate ideological foundations and proven successes.

The global public should be proud of the vast improvements made to the health of their environment in the last 30 years. The environmental fad of 1960s and 70s appears to have lasted longer than many would have predicted, to become forever a part of our culture. Indeed, the global environmental consciousness continues to be raised, and an improvement in the global quality of life is a worthy challenge for all nations. Scientific discovery and education is, and shall always be, the cornerstone for prosperity, cultural advancement and prudent natural resource management in free democratic societies.

So there you have it, environmental issues presciently recognized by President Teddy Roosevelt in the early 20th Century led to land conservation measures to control the irresponsible use of our natural resources. In the last half of the 20th Century, governmental agencies at all levels imposed vast regulatory systems to control virtually every human, commercial and industrial activity in the world—often without conclusive scientific evidence of any significant impact of the activities upon ecology or human health. Environmental issues where elevated from local to national to global significance, leading to national and international environmental regulations. In the last decade of the 20th Century, environmentalists attempted to globalize environmental regulations via the vehicle of the global warming issue. The globalization did not work, nor did a critical mass of worldwide political consensus materialize. Yet now, in the 21st Century, it has.

Environmental regulations have spawned not only the most comprehensive (and some would argue the most expensive) governmental involvement in every endeavor of human life and commerce, but also have propagated a succession of environmentally co-dependent enterprises; from the growth of multi-level government regulators and their agencies, to academic curricula in environmental studies, to environmentalist activism, to the green market place. Sound environmental policy must rest on the firm foundation of sound current science. We cannot solve problems of basic human needs without stable, growing economies. Democratic free enterprise and increased global economic integration provide the only proven opportunity to improve the condition of all humanity, make economies more efficient, and protect the environment.

Unfortunately, in ecopolitics, the driving force for new and expanded environmental regulations is political expediency (correctness) rather than good science applied to problem solving. There are many stunning successes from the environmental controls that were enacted and enforced over the last 30 years. You and I can be quite proud of the successes because we participated in the public, political will that brought environmental regulations. In addition, we have paid its enormous economic costs that are embedded in every product, service and activity we enjoy today. We must attribute a significant part of our present, nearly two decades, of continuous global economic growth and prosperity to the global assimilation, and market adjustments, to the ubiquitous economic costs of environmental regulations. The brief history of environmental regulations proves unequivocally that environmental protection can only be supported in democratic and prosperous economies. Environmental protection can not be a priority in third world or developing countries. They simply cannot afford it as a priority.

References

1. BBC News, May 2007.

2. Numbers Watch, United Kingdom, December 2007.

3. *Exxon Lamp*, Winter 2000.

4. US EPA Proposed Budget, January 2008.

5. *Scientific Method*, Michael James, May 2003.

6. *The Limits to Science*, Duane H. Fickeisen, June 2000.

7. *The Economist*, February 2008.

8. Eco America Website, January 2008.

9. *Green Gone Wrong: Ecopolitics Exposed*, Paul Taylor, January 2001.

10. *New York Times,* Andrew Revkin, January 2007.

11. *Opinion Journal*, Richard S. Lindzen, July 2006.

12. *Washington Post*, Bjorn Lomborg, October 2007.

13. *The Economist,* January 2008.

14. *Global Warming: The Origin and Nature of the Alleged Scientific Consensus*, Richard S. Lindzen, April 1992.

15. *The Hindu,* January 2008.

16. Associated Press, June 2007.

17. *The Wall Street Journal*, August 31, 1999.

18. *Investors Business Daily*, September 17, 2007.

19. *The Wall Street Journal*, August 22, 1999.

20. *Investors Business Daily*, July 30, 2007.

21. *The Economist*, May 19, 2007.

22. National Commission on Energy Policy, December 2007.

23. *Gilder Technology Report*, George Gilder, May 6, 1999.

24. *Atlas Economic Research Foundation*, Alejandro Chatuen, December 6, 1996.

25. *Investors Business Daily*, December 18, 2006.

26. *Investors Business Daily*, February 2008.

27. US Environmental Protection Agency Website, February 2008.

28. *The Economist*, December 1, 2007.

29. *Investors Business Daily*, November 26, 2007.

30. *International Herald Tribune*, Jeff Jacoby, August 21, 2007.

31. Pacific Research Institute, Greg Easterbrook, April 20, 2000.

32. *The Economist*, September 8, 2007.

33. Hudson Institute, Michael Fumento, April 2, 1999.

34. American Council on Science and Health, Elizabeth M. Whelan, July 27, 1999.

35. United Nations Website, February 2008.

36. *The Dictionary of Ecology and Environmental Science*, by Henry W. Art, Henry Holt and Co., 1993.

37. *Webster's New World College Dictionary*, Fourth Edition, 2000.

38. *The Economist*, November 4, 1995.

39. *The Culture of Fear*, Barry Glassner, Basic Books, 1999.

40. *Illness as Metaphor*, Farrar, Straus & Giroux, 1989.

41. *The Wall Street Journal*, March 15, 2000.

42. *The Prince of Tennessee*, Maraniss and Nakashima, Simon & Schuster, 2000.

43. Patrick J. Michaels, *USA Today*, February 2, 2007.

44. Ron Scherer, *The Christian Science Monitor*, October 13, 2006.

978-0-595-50152-6
0-595-50152-4